STARK LIBRARY JUN 2021

MEASURING AREA

by Meg Gaertner

Cody Koala

An Imprint of Pop!
popbooksonline.com

abdobooks.com
Published by Pop!, a division of ABDO, PO Box 398166, Minneapolis, Minnesota 55439. Copyright © 2020 by POP, LLC. International copyrights reserved in all countries. No part of this book may be reproduced in any form without written permission from the publisher. Pop!™ is a trademark and logo of POP, LLC.

Printed in the United States of America, North Mankato, Minnesota

102019
012020

THIS BOOK CONTAINS RECYCLED MATERIALS

Cover Photo: iStockphoto
Interior Photos: iStockphoto, 1, 15; Shutterstock Images, 5, 7, 9, 11, 13, 17, 19, 21
Editor: Meg Gaertner
Series Designer: Jake Slavik

Library of Congress Control Number: 2019942412
Publisher's Cataloging-in-Publication Data
Names: Gaertner, Meg, author.
Title: Measuring area / by Meg Gaertner
Description: Minneapolis, Minnesota : Pop!, 2020 | Series: Let's measure | Includes online resources and index.
Identifiers: ISBN 9781532165535 (lib. bdg.) | ISBN 9781532166853 (ebook)
Subjects: LCSH: Size and shape--Juvenile literature. | Size perception--Juvenile literature. | Measurement--Juvenile literature. | Mathematics--Juvenile literature. | Formulas (Mathematics)--Juvenile literature.
Classification: DDC 530.813--dc23

Hello! My name is
Cody Koala

Pop open this book and you'll find QR codes like this one, loaded with information, so you can learn even more!

Scan this code* and others like it while you read, or visit the website below to make this book pop.

popbooksonline.com/measuring-area

*Scanning QR codes requires a web-enabled smart device with a QR code reader app and a camera.

Table of Contents

Chapter 1
What Is Area?. 4

Chapter 2
Units of Measurement 6

Chapter 3
How to Measure. 12

Chapter 4
Measure It! 20

Making Connections 22
Glossary. 23
Index 24
Online Resources 24

Chapter 1

What Is Area?

Area is the amount of space in a flat shape. For example, the area of a wall is the size of its surface. Area tells us how many tiles would be needed to cover the wall.

Watch a video here!

Chapter 2

Units of Measurement

People can measure area by looking at a **grid**. All of the squares in a grid are the same size. They all have a side **length** of 1.

Grid

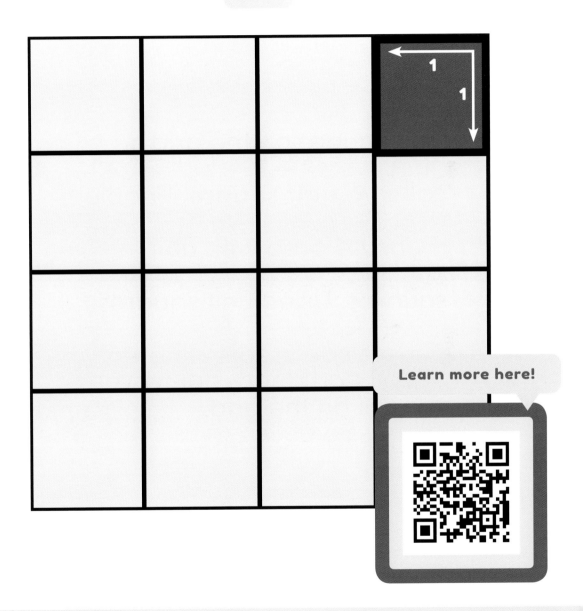

Learn more here!

A square on the grid is called a **unit square**. People measure area with unit squares. Using unit squares, people can create different shapes on the grid.

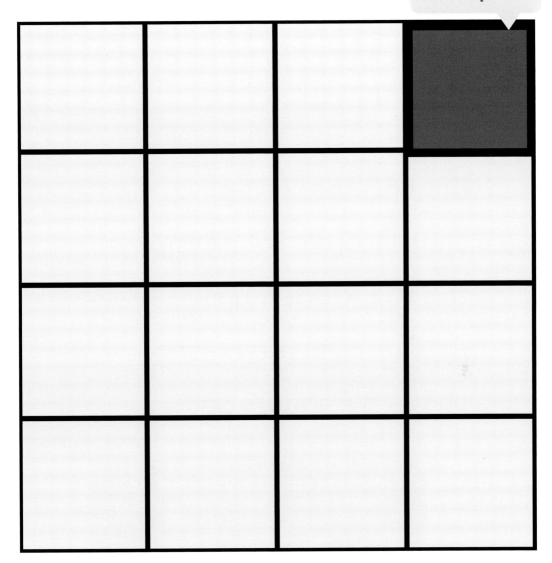

People count the number of unit squares in the shape. That number gives the shape's area. For example, this shape has 6 unit squares. The area of the shape is 6 square units.

> Make sure the unit squares cover the whole surface of the shape. There should not be any gaps.

	1	2	3
	4	5	6

Chapter 3

How to Measure

Squares and rectangles have a **length** and a **width**. People can find the area of these shapes by **multiplying** the length and the width.

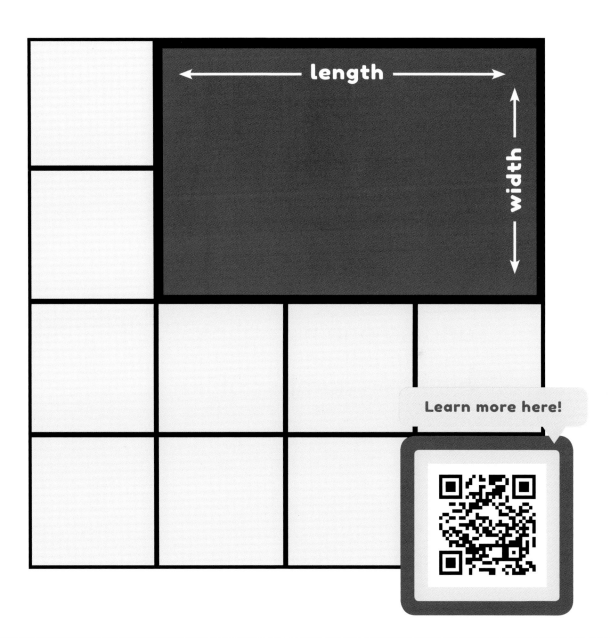

Multiplying is similar to adding. Multiplying involves adding one number to itself a certain number of times.

The first number in a problem tells what number to add. The second number tells how many times to add it. For example, multiply 3 and 2. To do this, add 3 to itself 2 times.

3 × 2 = 6
3 + 3 = 6

Multiply the length and width of this rectangle. The answer is the shape's area. It tells the number of **unit squares** in the shape.

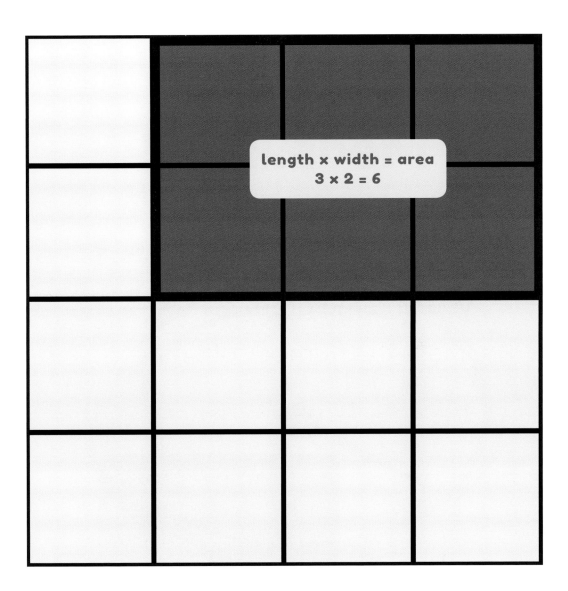

Chapter 4

Measure It!

Here is a new shape. Can you find the area?

> Sometimes a shape does not perfectly follow the **grid** lines. In this case, add the parts of a **unit square** together.

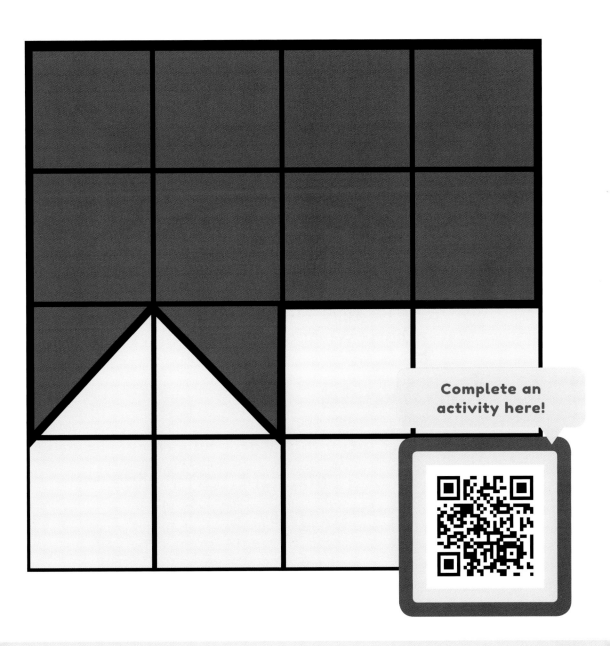

Complete an activity here!

Making Connections

Text-to-Self

Have you ever seen someone measure area? What shape was the area?

Text-to-Text

Have you read other books about measuring the size of something? What did you learn?

Text-to-World

Miguel wants to know the area of his desk. His desk is 5 feet (1.5 m) long and 3 feet (0.9 m) wide. What is the area?

Glossary

grid – lines that cross one another to form a series of squares.

length – the measurement of an object's longer side.

multiply – to add a number to itself a certain number of times.

unit square – a square with a side length of 1.

width – the measurement of an object's shorter side.

Index

adding, 14, 16, 20

grid, 6, 7, 8, 20

length, 6, 12, 13, 18, 19

multiplying, 12, 14, 16, 18

unit square, 8, 9, 10, 18, 20

width, 12, 13, 18, 19

Online Resources

popbooksonline.com

Thanks for reading this Cody Koala book!

Scan this code* and others like it in this book, or visit the website below to make this book pop!

popbooksonline.com/measuring-area

*Scanning QR codes requires a web-enabled smart device with a QR code reader app and a camera.